T0392371
SCIENCE

For all the daydreamers and thinkers:
The world needs you now more than ever. —KWL

For Mr. Gibson and those who encourage curiosity. —CL

The author would like to thank Anna Thanukos
of the University of California Museum of Paleontology at Berkeley
for her expertise and consultation.

ABOUT THIS BOOK

The illustrations for this book were created using a combination of traditional and digital media. This book was edited by Lisa Yoskowitz and designed by Angelie Yap. The production was supervised by Jonathan Lopes, and the production editor was Esther Reisberg. The text was set in Readex Pro, and the display type is Ideal Sans.

 • Little, Brown and Company • Hachette Book Group • 1290 Avenue of the Americas, New York, NY 10104 • Visit us at LBYR.com • First Edition: September 2025 • Little, Brown and Company is a division of Hachette Book Group, Inc. • The Little, Brown name and logo are registered trademarks of Hachette Book Group, Inc. • The publisher is not responsible for websites (or their content) that are not owned by the publisher. • Line from Albert Einstein's letter to his biographer Carl Seelig on p. 33 (*"I have no special talents. I am only passionately curious."*) from "To Carl Seelig, March 11, 1952," Albert Einstein Archives 039-013, Hebrew University of Jerusalem, reprinted with permission of Princeton University Press. • Little, Brown and Company books may be purchased in bulk for business, educational, or promotional use. For information, please contact your local bookseller or the Hachette Book Group Special Markets Department at special.markets@hbgusa.com. • Library of Congress Cataloging-in-Publication Data • Names: Larson, Kirsten W., author. | Li, Cornelia, illustrator. • Title: This is how you know / by Kirsten W. Larson ; illustrated by Cornelia Li. • Description: First edition. | New York : Little, Brown and Company, 2025. | Includes bibliographical references. | Audience: Ages 4–8 | Summary: "A picture book introduction to scientific inquiry, celebrating the power of curiosity and science to solve big problems and make the world a better place." — Provided by publisher. • Identifiers: LCCN 2024007758 | ISBN 9780316283281 (hardcover) | ISBN 9780316283380 (ebook) • Subjects: LCSH: Science—Philosophy—Juvenile literature. | Reason—Juvenile literature. • Classification: LCC Q175.2 .L36 2025 | DDC 501—dc23/eng/20240522 • LC record available at https://lccn.loc.gov/202400775 • ISBN 978-0-316-28328-1 • PRINTED IN DONGGUAN, CHINA • APS, 5/25 • 10 9 8 7 6 5 4 3 2 1

THIS IS HOW YOU KNOW

By Kirsten W. Larson
Illustrated by Cornelia Li

LB
Little, Brown and Company
New York Boston

You KNOW . . .

Earth warms up,
brushed by Sun's gentle rays.

Soil cradles tomatoes,
which sprout up
with water and light.

Your swing soars up
with the pump of your legs—*whoosh*—
or a push from a friend,
sending you high.

All of this is science . . .
which is *how* you know.
It's the way you make discoveries,
starting with natural curiosity:

watching and wondering,

reading and tinkering,

daydreaming and thinking.

Until you uncover a question
you can't answer.
Not yet.

That question
becomes your guiding star in the dark.

Read,
watch,
and wonder.
What could be the answer?

Come up with an idea,
then let your imagination lead you:
What would you see
if your idea were true?

Next, search the world for clues,
tiny glimpses of truth.

Look in labs.

On slides or screens.

Between the trees.

Under leaves.

Among the stars.

Don’t just gather one clue or two.
Heap them up a mile high.

Look for patterns and connections that matter
on your winding path to a breakthrough.

Remember, though, it won't be easy.

You may change course
or change your mind.
Or you might realize your idea was wrong all along.
Then you must begin again.

That's okay.
It's the way science works—
it's surprising,
messy,
beautiful.

And often,
knowing an answer is wrong
is as good as finding out
what's right.

So just keep going,
spending hours combing
through the ordinary,
hoping to find the extraordinary.

And when you do . . .

EUREKA!

Something new!
And you're the first to know.

But that's not the end
of your adventure.

Not yet.

Because every discovery
is a chance
to chart a new course.

With the next question on your lips.
Wonder beating in your heart.
The greatest joy.

Because this is science—
and science needs *you*.

It needs all of us
working together.
In every corner of
the world.

Sharing our curiosity.

Comparing our discoveries.

Solving big problems.

Making our world
a better place.

Uncovering whole new worlds,
places we've seen only in dreams.

Because this is science,
which starts with the
natural world and noticing.

It’s sparked by imagination
but shaped by hard work.

And there’s always
more to KNOW.

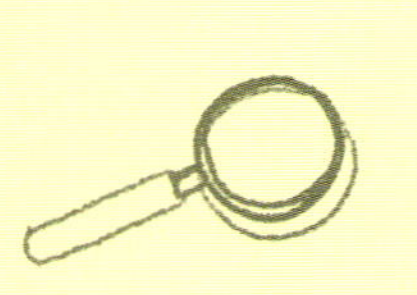

How Science Works

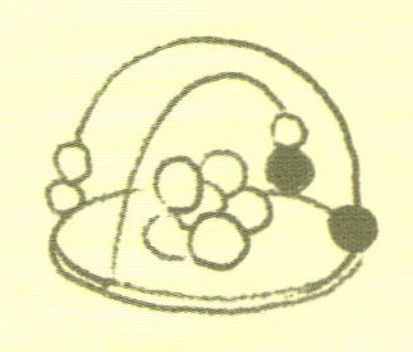
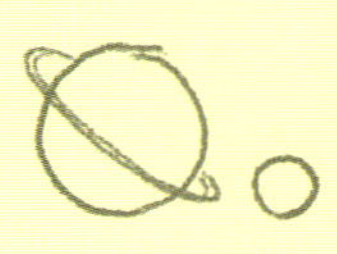

Science is not just a subject in school; it's a process too. That process, known as scientific inquiry, is the way we study the natural world, making new discoveries every day.

All scientists engage in scientific inquiry, even if their journeys vary. Some start with a question in mind. To test their idea, they do lab experiments or build a computer model or simulation. Others start by gathering clues, staring at stars, digging up fossils, or collecting butterflies. Then they look for patterns or something that seems odd, which sparks an idea. There's no single straight path to follow when it comes to scientific inquiry. Think of it more like a dance than a race to the finish line.

Starting with natural curiosity. This is where all science begins, with wondering, noticing, paying attention to things that seem interesting or different. Does climbing trees make your heart sing? Could you stare at clouds for hours? Do you like to race cars down tall ramps? Any of these could be a starting point for science.

Until you uncover a question you can't answer. Not yet. Being curious means asking a lot of questions. You may wonder: *How do plants drink water? When clouds stack up high like cauliflower, will it rain soon? What would make my race car go faster?*

Scientists often focus on questions that nobody knows the answers to. That's where discoveries lie. What's your question?

That question becomes your guiding star in the dark. You may already have a few ideas about the answer to your question. Or you may have to do some more research first. But your idea about what's happening, called your hypothesis, is at the heart of scientific inquiry. It is what you'll test by collecting clues or observations.

Your idea may be that plants have strawlike parts to suck up water from the ground. Or that high, puffy clouds bring rain. Or maybe you have an idea that a taller ramp will send your race car farther. What hypothesis may answer your question?

What would you see if your idea were true? It's almost time for the fun part: experimenting, building models, using test tubes and magnifying glasses. But before you start, think carefully: If your hypothesis were true, what would you expect to see? What kind of results would you expect to get? Framing your expectations, also called predictions, as an if-then statement can help.

So, if trees and plants suck up water like a straw, then celery stalks or cabbage leaves placed in red water should slurp up the water and turn red. If puffy clouds bring rain, then keeping a weather chart should show periods of rain following sightings of puffy clouds. And if a ramp's height determines how far a car goes, a taller ramp should send your car farther across the floor.

This step is important, because expectations give you something to compare your evidence to. If they match, your guess may be right. If they don't, you may need to test out a different answer.

Next, search the world for clues. It's time to carry out your test and collect clues to the truth. These clues are also called data, observations, or evidence. Put those celery stalks and cabbage leaves in cups of colored water, and watch what happens. Make a chart to track the cloud shapes and weather you see each day. Build ramps, race your cars, and measure how far they go. Record your results. Go for it! Collect lots and lots of evidence.

Look for patterns and connections that matter. If you're drowning in data, math is like your life preserver. Math helps us make sense of information and see patterns and relationships. You might use graphs or charts to help organize your results. For example, a graph could show the relationship between a ramp's height and how far the car goes.

Eureka! Did your observations match your expectations? Do your results support your idea? Fantastic! You've had a breakthrough.

But if not, that's okay too. Because ruling out one hypothesis is a step along the path. Scientists often test many explanations before they find the one that fits the data. When things don't work out the first time, don't give up. Rethink your hypothesis and predictions, and try again. It's all part of the process.

Science needs you. It needs all of us working together. Scientists don't work alone. They share their results and explanations with others. They might write up their discoveries in scientific journals or make presentations. Other scientists review the discoveries and even run the same experiments to see if they get the same results.

Being a scientist means staying open-minded. Science is always open to change and new ideas, but only after those new ideas have been fully tested by many people and are supported with lots of evidence. Working together, scientists all over the world add to the body of knowledge we call science.

Solving big problems. Making our world a better place. Scientists help solve big, real-world problems. Their research gives people in government ideas for how to stop diseases from spreading, protect endangered animals, save our warming planet, and so on.

And there's always more to know. The world always needs more scientists—people with passion, curiosity, and the drive to comb through the ordinary, looking for the extraordinary. People like YOU!

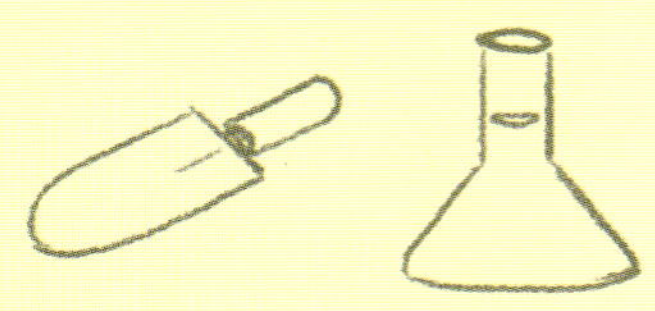

Learn More About Science

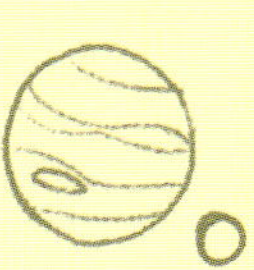

American Museum of Natural History, "Ology," https://www.amnh.org/explore/ology.

Beaty, Andrea. *Ada Twist, Scientist.* New York: Abrams Books for Young Readers, 2016.

Burns, Loree Griffin. *You're Invited to a Moth Ball: A Nighttime Insect Celebration.* Watertown: Charlesbridge, 2020.

Girl Scouts of the United States of America, "STEM," https://www.girlscouts.org/en/raising-girls/school/STEM.html.

Harrington, Janice N. *Buzzing With Questions: The Inquisitive Mind of Charles Henry Turner.* Honesdale: Calkins Creek, 2019.

NASA, "NASA STEM Engagement," http://www.nasa.gov/stem.

Offill, Jenny. *11 Experiments That Failed.* New York: Schwartz & Wade, 2011.

University of California Museum of Paleontology, "Understanding Science: How Science *Really* Works," https://undsci.berkeley.edu/index.php.

What Scientists Say About Science

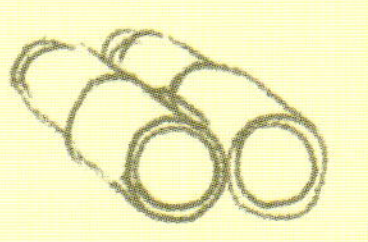

"The very nature of science is discoveries, and the best of those discoveries are the ones you don't expect."

—Neil deGrasse Tyson, astrophysicist, 1958–

"We especially need imagination in science. It is not all mathematics, nor all logic, but it is somewhat beauty and poetry."

—Maria Mitchell, astronomer, 1818–1889

"Progress is made by trial and failure; the failures are generally a hundred times more numerous than the successes."

—William Ramsay, chemist, 1852–1916

"An experiment is a question which science poses to Nature, and a measurement is the recording of Nature's answer."

—Max Planck, physicist, 1858–1947

"I am among those who think that science has great beauty. A scientist in his laboratory is not only a technician: he is also a child placed before natural phenomena which impress him like a fairy tale."

—Marie Curie, physicist and chemist, 1867–1934

"I have no special talents. I am only passionately curious."

—Albert Einstein, physicist, 1879–1955

"'Now you see how easy it is to understand.' 'So are all truths, once they are discovered; the point is in being able to discover them.'"

—Galileo Galilei, astronomer, 1564–1642